challenging
MATH

Glen Vecchione

Illustrated by Glen Vecchione
& Nina Zottoli

Sterling Publishing Company 10051419
New York

Edited with pages designed by Jeanette Green
Computer graphics by Nina Zottoli and Glen Vecchione
Cartoons by Glen Vecchione

Library of Congress Cataloging-in-Publication Data
Vecchione, Glen.
    Math challenges : puzzles, tricks & games / Glen Vecchione ; illustrated by
Glen Vecchione and Nina Zottoli.
        p.   cm.
    Includes index.
    ISBN 0-8069-8114-8
    1. Mathematical recreations. I. Title
QA95.V43   1997
793.7'4—dc20   96-43526

                        10 9 8 7 6 5 4 3

            First paperback edition published in 1998 by
                    Sterling Publishing Company, Inc.
            387 Park Avenue South, New York, N.Y. 10016
                Previously published as *Math Challenges*
                    © 1997 by Glen Vecchione
                Distributed in Canada by Sterling Publishing
        % Canadian Manda Group, One Atlantic Avenue, Suite 105
                    Toronto, Ontario, Canada M6K 3E7
            Distributed in Australia by Capricorn Link (Australia) Pty Ltd.
    P.O. Box 6651, Baulkham Hills, Business Centre, NSW 2153, Australia
                Manufactured in the United States of America

                    Sterling ISBN 0-8069-8114-8 Trade
                                0-8069-8115-6 Paper

Contents

For Briana Vecchione
Little Wizard!

Chapter 1

GRID & DOT GAMES

**Squiggling Snake • The Cop & the Robber •
Springing Sprouts • Hare & Hounds • The Ratcatcher
• Horse Race • Daisy Petals • Black & White**

All games are mathematical in that they present both a problem
and a method for solving the problem. Board games like checkers
or chess depend on the competing player's powers of calculation.
Games that use playing cards, number wheels, or dice combine
the player's skill with the laws of probability. Most games teach us
about what mathematicians call *quantum operations*. This means
that players must repeat small, uniform steps to reach a solution
within a given framework of laws. So, playing a game of checkers
isn't very different from solving a math puzzle.

SQUIGGLING SNAKE

In this game, players take turns joining dots by a line to make one long snake. No diagonal lines are allowed. Each player adds to the snake at either end, and a player can only add to his opponent's segment, not his own. The first player to make the snake close on itself loses the game.

Here's how an actual game might be played. To tell the two players apart, one player draws straight lines, and the other player draws squiggly lines.

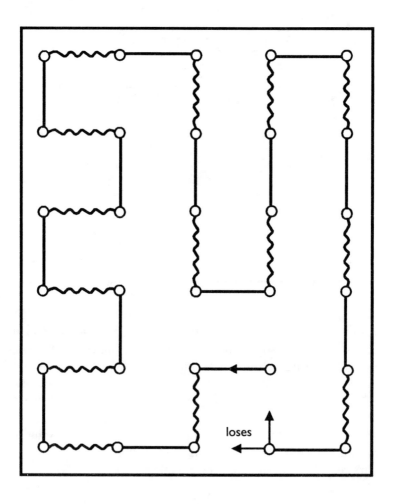

loses

THE COP & THE ROBBER

On a piece of cardboard or construction paper, draw the game board below including the letters C (cop) and R (robber). This board represents a city grid of several blocks and streets, and the letters indicate the starting position of the cop and robber.

You need two different coins, one for the cop and one for the robber. Start with each coin on its letter. The cop always moves first. After that, robber and cop take turns moving. A player moves a coin one block only, left or right, up or down—that is, from one corner to the next. The cop captures the robber by landing on the robber in one move. To make the game a little more challenging, the cop must capture the robber in twenty moves or less, or the robber wins.

Hint: There is a way for the cop to nab the robber. The secret lies in the bottom right corner of the grid.

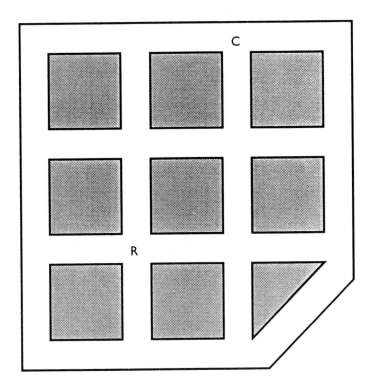

SPRINGING SPROUTS

Most mathematical games are played on grids, but some of the newer ones use *topology,* the geometry of flexible lines and surfaces, as a starting point. Springing Sprouts was invented in the 1960s by a mathematician at Cambridge University in England. Not only will you enjoy playing it, but your finished game makes an interesting piece of artwork!

The first player begins by drawing three spots (A). The second player must connect two of the spots with a line (remember, the line can be curved) and then adds a new spot somewhere along that line (B).

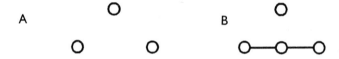

No lines may cross (C), but a player may connect a spot to itself in a loop—as long as he adds another spot (D).

A spot "dies" when three lines lead to it and no more lines can connect to it. To indicate a dead spot, darken it (E).

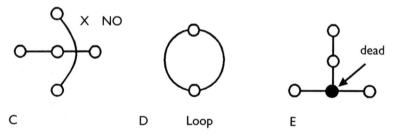

A player wins by drawing the last connecting line so that all the remaining spots are dead and the second player can no longer connect them. Here's a sample game, won by player A in seven moves.

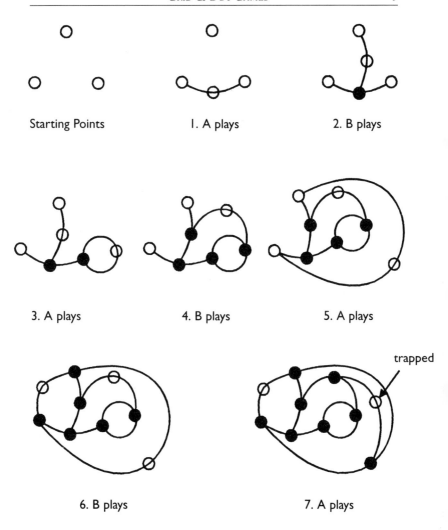

Starting Points 　　　1. A plays 　　　2. B plays

3. A plays 　　　4. B plays 　　　5. A plays

trapped

6. B plays 　　　7. A plays

Mathematicians have tried to figure out how many moves it takes for either player to win Springing Sprouts. They've discovered, but not yet proved, that the number lies between twice and three times the number of spots you start with. Starting with three spots, for instance, the game can continue for six to nine moves. Starting with four spots, the game may last for eight to twelve moves, and so on.

HARE & HOUNDS

Many board games involve "hunting." This one was a favorite of Victorian schoolchildren. Games like this teach us about geometrical figures called *trapezoids,* which are four-sided figures with only two sides parallel. The corners of trapezoids, or *vertices,* also play an important part in this game.

Use a ruler to draw the board on a piece of paper or cardboard. At every place where the lines meet in a corner, draw a small circle. Draw a large A at the left side of the board and a large B at the right side.

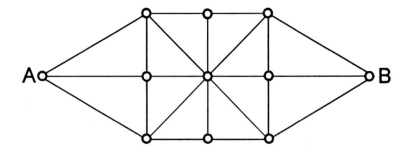

Player #1 has one coin, representing the Hare, and Player #2 has three coins, representing the Hounds. The game starts with Player #1 placing his Hare on circle A. Player #2 follows by placing one of this Hounds on any other circle. He will use the next two turns to place his other Hounds on circles, as the Hare moves from circle to circle, trying to escape.

The Hounds may move in any direction forward (that is, towards circle A) or up and down, but not towards circle B. The aim of the Hare is to reach the safety of circle B while the Hounds, of course, try to block his way and prevent him from moving.

THE RATCATCHER

This is another version of a strategic chase-and-capture game. One player has twelve Rats, and the other player is the Ratcatcher. The game starts with the thirteen pieces in the positions shown below, the white dot representing the Ratcatcher.

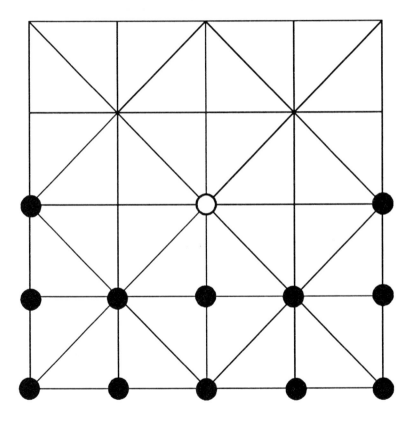

Both Ratcatcher and Rats can move in any direction onto an empty intersection of lines (*vertex*). The Ratcatcher can remove a Rat from the board by jumping over him to an empty vertex on the other side. But the Rats can gang up on the Ratcatcher by surrounding him in such a way that he can neither move nor jump.

HORSE RACE

In this ancient Korean racing game, *nyout,* players throw one of a pair of dice (a die) to determine the number of moves, then race their buttons, or "horses," around the circular track illustrated below. Notice that the track has five large squares, sixteen smaller squares, and eight circles.

The players start at square A, and the first player to return to square A wins the game. When a player throws the die, he may only move up to five spaces; a throw of six must be thrown again. If a player's horse lands on one of the larger squares B, C, or D, the horse may take a shortcut through the circle to reach A.

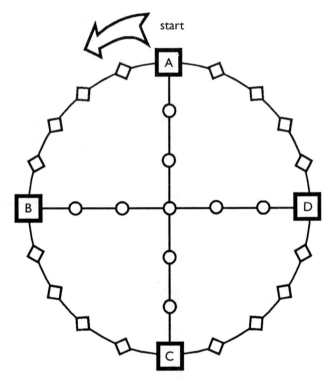

Each player may have up to three horses, but both players must have the same number of horses to ensure a fair game. A player may choose to add another horse instead of moving his original horse.

If one player's horse lands on a square already occupied by another of his own horses, both horses may move together in all the following moves. If one player's horse lands on a square occupied by his opponent's horse, the opponent's horse is removed and must reenter the race.

To win the game, a player must throw the exact number to get each of his horses back to A. It usually takes three or four times around the track before one player's horses win. And your fortunes may reverse at any time!

DAISY PETALS

On a piece of paper or cardboard, construct a thirteen-petal daisy using crayons for the petals.

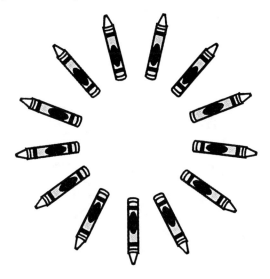

Two players take turns plucking either one petal apiece, or two neighboring petals apiece. The player who plucks the last petal wins the game.

Remember, a player can only take two petals if the petals are neighbors. The player who makes the second move can always win this game if he has a sharp eye and knows something about the principle of *symmetry*—that is, the balance of parts on opposite sides of a line or about a center point.

BLACK & WHITE

One player alone can enjoy this game, or two players can compete to see who can finish in the least number of moves.

Reproduce the board below with two 3-inch-square pieces of paper, joined at the corner and lined in a grid pattern. In squares on the left side of the board, place eight black pieces (buttons or pennies will do). In the squares on the right side of the board, place eight white pieces.

The object of the game is to exchange the positions of the black and white pieces in the least possible number of moves. You can move a piece by sliding it to a neighboring empty square, or by jumping over a neighboring piece of either color. This can be done in exactly 46 moves, but you can still consider yourself an expert if you finish in 52 moves or less.

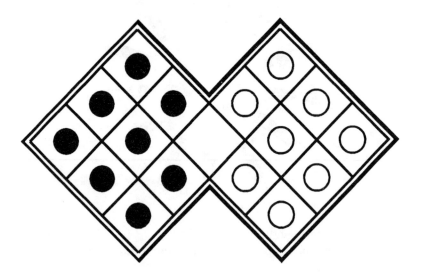

Chapter 2

SUM OF THE PARTS

Stamp Stumper ● Broken Dishes ● Cut the Pizza ●
Fractured Fractions ● Divide the Time ●
Parcels of Land ● Four Lines in a Square ●
Count the Blocks ● Sides, Edges & Corners ●
Crayon Constructions ● Box the Dots

Mathematicians have two ways of solving puzzles: piecing together small bits of information to understand larger problems and breaking down complicated ideas into simpler parts. For example, *geometry* helps them understand how to combine certain shapes to make larger shapes or how to reduce certain shapes into smaller shapes. *Fractions* help them understand the functions of whole numbers. In each case, breaking something down in order to put it back together again can lead to a valuable understanding of basic mathematical principles.

STAMP STUMPER

Start with a sheet of 24 stamps. Following the diagram below, tear out two sets of three stamps, making sure the stamps of each set remain joined. Tear out a third set, then a fourth set. How many sets, total, can you tear from the sheet?

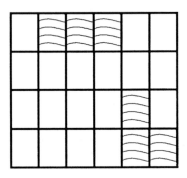

BROKEN DISHES

If you could put all the broken pieces back together in this drawing, how many dishes would you have?

CUT THE PIZZA

A group of eight people walked into a restaurant and ordered a large pizza. The place was very busy, so when the pizza arrived, the waiter wanted to cut it up into eight even pieces as quickly as possible. He did this with only three straight cuts of his knife. Can you figure out how the waiter cut the pizza?

FRACTURED FRACTIONS

In this puzzle, you must find two-thirds of three-fourths of five numbers.

Find 2/3 of 3/4 of 12.
Find 2/3 of 3/4 of 20.
Find 2/3 of 3/4 of 32.
Find 2/3 of 3/4 of 44.
Find 2/3 of 3/4 of 52.

Can you discover the trick to doing this quickly?

DIVIDE THE TIME

By drawing only two lines, divide the clockface below so that the numbers in each section add up to the same sum.

PARCELS OF LAND

A landowner died and left a large, square piece of land to his wife and four sons. His wife received one-fourth of the land (section A), and his sons had to parcel out the remaining three-fourths of land equally. Draw a picture showing how the landowner's sons divided the land. Remember, each of the four sections must be the same size and shape.

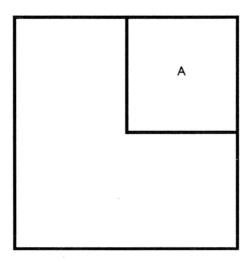

FOUR LINES IN A SQUARE

With a pencil and ruler, draw a square. Then draw four straight lines so that each line connects opposite sides of the square. Arrange your four lines so that you divide the square into as many sections as you can. Can you figure out the maximum number of sections you can make with only four lines?

COUNT THE BLOCKS

Count the number of blocks in each arrangement. Assume that visible blocks rest on identically shaped hidden blocks, and that every arrangement is solid.

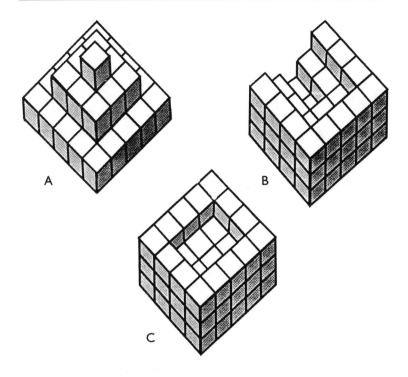

SIDES, EDGES & CORNERS

In the arrangement below, figure out how many blocks each numbered block touches. Blocks touch if any one of their sides, edges, or corners come in contact.

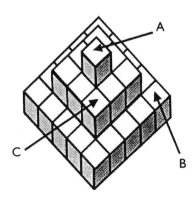

CRAYON CONSTRUCTIONS

With 24 crayons, construct a large square. How many crayons does each side of the square contain?

Construct two squares of the same size. How many crayons are there to each side?

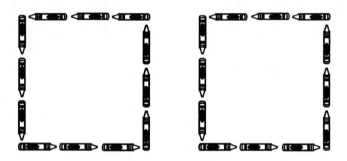

Now construct three squares of the same size. How many crayons are there to each side?

With one crayon to each side, you can make six identical small squares as shown.

1. How do you determine how many squares of the same size you can construct with 24 crayons?

2. Now construct squares of different sizes, using 24 crayons.

A. With three crayons to a side, how many smaller squares do you get?

B. Construct squares with two crayons, maximum, to a side to make seven squares of two different sizes.

C. Construct seven *identical* squares with one crayon to a side.

D. Also, with one crayon to a side, construct eight and nine identical squares. How many larger squares does each design contain?

Clue: Squares can overlap and there'll be some bigger squares containing smaller ones.

BOX THE DOTS

Divide the hexagon below so that each dot is in its own rectangular "box." All the boxes must be the same size, and there should not be any spaces between the boxes.

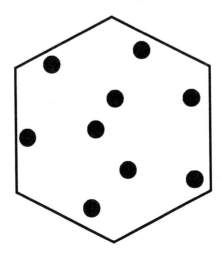

Hint: Think of the hexagon as a three-dimensional object.

Chapter 3

VANISHING TRICKS

**The Lost Line ● Disappearing Square ●
Where's the Wabbit? ● The Floating Hat ●
Lions & Hunters ● Funny Money ●
Mysterious Tangrams ● Tangram Oddities**

These tricks are all based on the mathematical principle of *concealed distribution*. This complicated-sounding term just means that geometric shapes can be broken up and recombined so that they are completely different, yet the same. Sound impossible?

THE LOST LINE

Start with an index card. Using a ruler, draw ten lines on the card, making sure that each line is exactly the same length. Next, use the ruler to trace a diagonal line from a top corner of the card to a bottom corner. Cut the card in half along this line.

Now slide one half of the card along the other half just far enough so that the ten lines turn into nine lines. Where did the tenth line go?

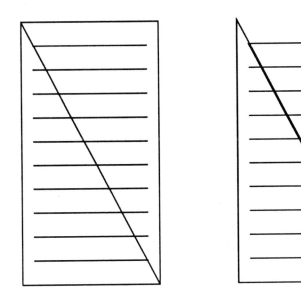

DISAPPEARING SQUARE

Using the same principle, you can also make a square vanish. Start with a 4 × 4-inch piece of paper that you have measured into sixteen 1-inch squares. Trace a diagonal line that goes from the top right corner of the fourth square, top row, to the bottom left corner of the second square, bottom row.

Notice that this diagonal line does not divide the paper from corner-to-corner, and it does not even touch the corners of the three interior squares. But this is the secret of the trick.

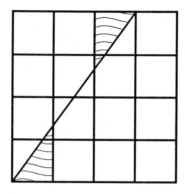

Cut along the diagonal line, dividing the paper. Put the paper back together and count the small squares again. Now, slide the right part of the paper down, lift it, then place it slightly over the left side of the paper until you have fourteen squares and two "half-squares" at the top right and bottom left. Snip off the half-square at the top right corner and fit it into the space at the bottom left corner. Now you have fifteen squares. The sixteenth square has disappeared.

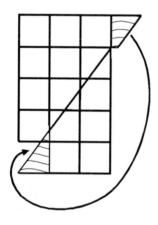
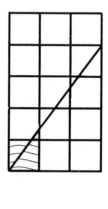

Now you have a 3 × 5-inch rectangular piece of paper containing fifteen squares, when you started with a 4 × 4-inch square piece of paper containing sixteen squares. Where is the missing square?

WHERE'S THE WABBIT?

Following the illustration below, draw two intersecting lines that divide a rectangular piece of paper. Label them A, B, and C as shown.

Carefully draw eleven rabbits down the center line, following the illustration. Your rabbits don't have to be perfect—and they don't even have to be rabbits—but something with big ears and a big tail helps the trick's effect.

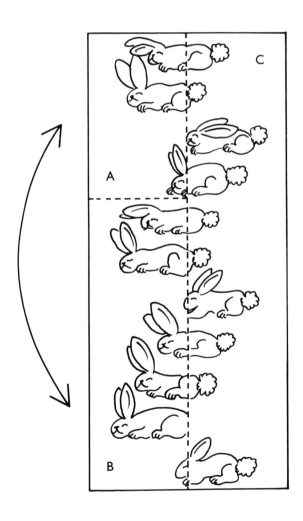

After you finish your rabbits, carefully cut along the lines you drew earlier, dividing the paper, and rabbits, into sections. Now switch the places of sections A and B against section C so that a whole new set of rabbits appears. How many rabbits?

THE FLOATING HAT

This example of the mathematical principle of concealed distribution has a mysterious touch.

Trace a line down the center of a strip of paper as shown in the illustration. Draw six heads, each wearing a funny hat. Make sure you draw each head according to the illustration—that is, draw the first head completely above the center line, cutting off the chin slightly. Draw the second head so that the line is right below the mustache, draw the third head so that the line divides the nose, and so on.

Cut along the line you traced. Slide the upper strip to the right along the lower strip until the heads recombine to form new heads. What happens?

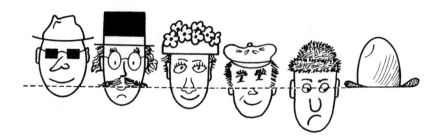

LIONS & HUNTERS

The 19th century American puzzlemaker Sam Loyd invented a concealed distribution puzzle that is perhaps the most puzzling of all. Notice the seven lions and seven hunters arranged around the inside and outside of the circle.

Photocopy the drawing three times and carefully cut out the circle from copy #3. Using the first copy as a guide, place the cutout circle from copy #3 over the circle in copy #2, completely covering it. Count the number of lions and hunters again.

Slowly rotate the cutout circle to the left so that the original set of lions and hunters disappears and a new set forms. Count the number of lions and hunters, comparing this new picture to your original one. Now, who do you think has a better chance in this hunt—lion or hunter?

FUNNY MONEY

In this trick, you can reverse the process of concealed distribution and make something appear seemingly out of nowhere. The trick was used by swindlers to make, say, fifty-one dollar bills out of fifty. The idea was that if the area lost in each bill were small enough, it would not be noticed and an extra bill would appear. You can try this for yourself using six play money bills or seven photocopies of a single play-money bill. Do not photocopy or cut up real money.

Arrange the seven bills in a column, like the illustration. Use a ruler and pencil to trace a line along the left edge of each bill, making each line a little further to the right than the line on the previous bill. You will end up with seven bills, each divided differently into unequal sections (A).

Cut each bill, following the line. Keep the cut bills in the same column arrangement you started with.

A   B

Now, move all the cut pieces on the left down the column by one bill, and tape those pieces to the pieces on the right. You will wind up with a new column of eight bills (B). Notice how the top and bottom bills of the column consist of uncut pieces. Where did the extra bill come from?

MYSTERIOUS TANGRAMS

The *tangram,* or square puzzle, originated in China over 4,000 years ago and became very popular in Europe. Lewis Carroll enjoyed solving them. So did Napoleon, who even invented a few of his own.

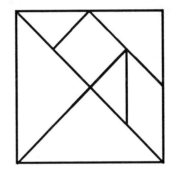

Original Tangram

Tangrams show how geometric shapes can be broken up and recombined to form new shapes. Some of these new shapes contain surprises. The diagram shows the seven pieces that make up the tangram. Use a ruler and marker to trace these pieces on a square of cardboard, then cut them out.

The challenge is to take these seven pieces and recombine them into various pictures or figures. See if you can figure out how to construct these animals using all seven pieces. You can flop pieces.

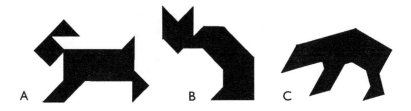

A   B   C

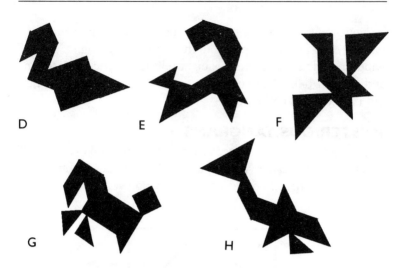

D   E   F

G   H

TANGRAM ODDITIES

Now, for some oddities. Notice the two tangram pigs below. Each pig uses all seven pieces, and yet only one pig has a tail. If each tangram contains all seven pieces, then how can the tangram on the right appear to have one extra piece?

This is also true for the two figures below. How is it that only the figure on the right has a foot when each figure uses all seven pieces? Where did the extra foot piece come from?

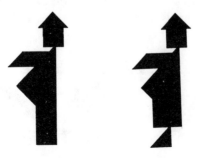

Chapter 4

STRETCHY SHAPES & SQUIGGLY LINES

**Impossible Postcard ● Möbius Mysteries ●
Squaring the Loops ● Two Knotty Problems ●
A Fly's View ● Power Lines ● The Bridges of Königsberg**

We live in a world of many shapes that are much more interesting than those found in a draftman's flat and angular drawings. A special branch of geometry, called *topology*, shows what happens when shapes change into different shapes. Topology also recognizes the importance of a shape's interior and exterior regions and how they relate to problem-solving. Topological puzzles often call for creative solutions and can demonstrate unusual and surprising principles of spatial relationships.

IMPOSSIBLE POSTCARD

Do you think you can cut a hole in a 3 X 5-inch postcard so that you can fit your head through the hole without tearing the postcard? You don't have to try this to know it doesn't work, since you can't cut a hole larger than the edges of the postcard.

But it does work. The trick is in cutting the postcard a certain way and then stretching it so that the edges of the postcard remain while you pass your head through the hole in its center. Here's how you do it.

1. Fold a postcard in half the long way. 2. Cut it with scissors as shown. 3. Then cut from A to B.

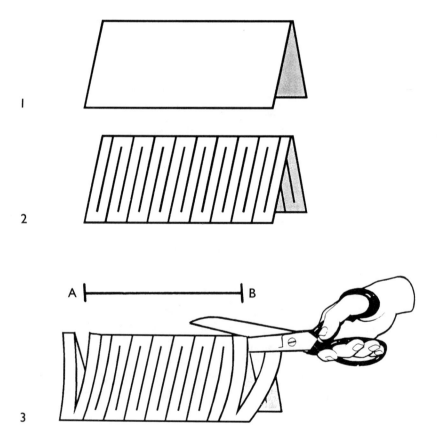

Finally, unfold the postcard and gently pull it apart at the narrow edges. You can see that the postcard easily fits over your head, if not your entire body.

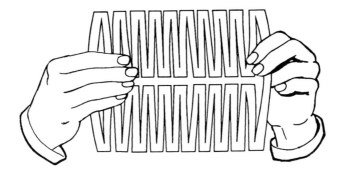

This demonstrates how a topologist would solve the problem. Topologists are interested in the qualities of a shape that can be changed (*variants*) and the qualities of a shape that remain the same (*invariants*). Although you can hardly recognize your original postcard as you stretch it over your head, the postcard still has edges and vertices that become visible again if you fold the postcard back to its original form.

MÖBIUS MYSTERIES

When you take and connect the ends of a strip of paper, you make a simple loop with two sides, two edges, an inside and an outside. If you draw a line around the outside of the loop, the line joins with itself and remains on the outside. If you draw a line around the inside of the loop, the line joins with itself and remains on the inside. The line drawn on the inside never meets the line drawn on the outside and vice versa.

But you can make a different kind, a Möbius loop (it's also called a Möbius strip, named after the 19th century German mathematician August Möbius), by taking a strip of paper and giving one end a half-twist before you join it to the other end. This way, the top of the strip becomes the bottom, and the bottom of the strip becomes the top. And you'll have a strange loop with only *one side and one edge.* To prove it, draw a line around the loop. Eventually, your line goes completely around and meets itself.

Ordinary Loop

Möbius Loop

Imagine two strips of paper sandwiched together and twisted into a Möbius loop. Now imagine a bug crawling between the strips of paper. He could crawl around the loop forever with a "ceiling" above him and a "floor" below without knowing that the ceiling and floor were actually the same surface. If he drew a big X on the floor and continued crawling, he would eventually reach his X again—on the ceiling. These are some of the reasons Möbius loops fascinate topologists.

Just as we've transformed other shapes, we can now change a Möbius loop into something quite different.

Cut two strips of paper, but before you connect their ends, draw a dotted line along the lengths of each strip. Connect the ends of one strip so that you have an ordinary loop, and connect the ends of the other strip so that you have a Möbius loop. Use a small amount of glue to connect the ends of your loops, so that they don't come apart when you cut them.

dotted lines   **dotted lines**

Ordinary Loop   **Möbius Loop**

What do you think will happen if you cut each loop along the line? First, cut along the dotted line on the ordinary loop. You can see that the cut divides the loop into two loops, as you might expect. Next cut along the line on the Möbius loop. Although you might expect to make two separate loops, you wind up with one long loop with two twists.

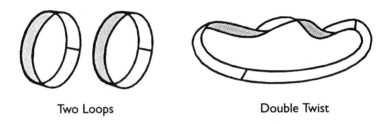

Two Loops   Double Twist

Deluxe Möbius Loop

Take another strip of paper and draw a line along its length as before. This time, give one end a complete twist before you attach it to the other end. Your new Möbius loop should have a smaller loop inside.

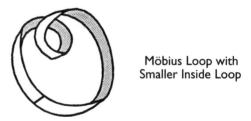

Möbius Loop with
Smaller Inside Loop

Cut along this line as before. This time you'll wind up with two twisted loops linked together.

Deluxe Möbius Loop

Super Deluxe Möbius Loop

Finally, make a Möbius loop where you give one end one and a half twists before joining it to the other end. Cut this strip along the line you drew, as before. You'll wind up with something completely different this time: a single loop linked together three times.

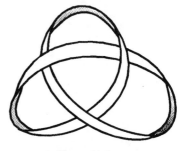

One and a Half Twists   Three Links

Try looping strips together in different ways and cutting them apart. See what strange topological results you can discover. No wonder many mathematicians choose to become topologists. It allows them to play with Möbius loops!

SQUARING THE LOOPS

Topologists study how shapes can change into different shapes while keeping some of their original qualities. They call the changing qualities of a shape *variants* and the unchanging qualities of a shape *invariants*. Unlike geometry, topology views a circle and a square as two different forms of the same object. Although this kind of thinking may seem a little difficult to understand at first, the following project will help make it clear.

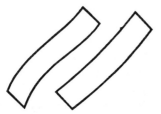
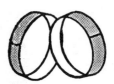

Take two strips of paper of equal length and width, and draw a straight line on each lengthwise. Connect the ends of each strip so that you have two simple loops. Instead of tape, use a small amount of glue to join your loops, so that they don't come apart when you cut them.

Glue one loop to the outside of the other loop, according to the illustration.

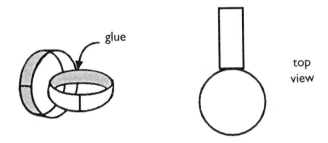

After the glue dries, carefully cut each loop along the line you drew earlier. As you continue to cut and your connected loops fall apart, you'll begin to notice something strange happening. When you finish cutting, carefully unfold this new object—a perfect square.

TWO KNOTTY PROBLEMS

By changing shapes into other shapes, you can get out of some tricky predicaments. Take knots, for example, sometimes what seems like a tight situation just requires some sleight of hand and a little topological know-how.

Quick Escape

Connect your wrists with a piece of rope about 14 inches (35 cm) long. Make sure the loops around your wrists are comfortable and not too tight. Have a friend begin to do the same, but before your friend ties his other wrist, loop his rope around yours (see illustration).

Try to separate yourself from your friend without untying the knots or cutting the rope. Although it might seem impossible at first, you can easily escape from your friend by slipping the rope loop over one of your hands and then back under one of the loops around your friend's wrist.

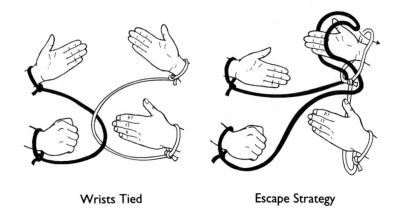

Wrists Tied   **Escape Strategy**

Freeing the Mug

Double up a long piece of string, loop it through the handle of a cup or mug, and tie the ends of the string together on a rod.

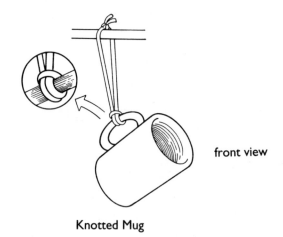

front view

Knotted Mug

Although the string appears knotted around the mug's handle, you can easily free the mug. First, loosen the loop at the handle. Then, widen the loop by pulling more string through it. And finally, pull the mug through the widened loop.

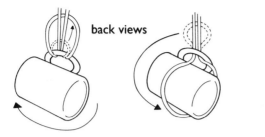

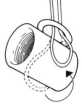

1. Loosen loop   2. Widen loop.   3. Pull cup free.

A FLY'S VIEW

A topologist views geometrical figures a little differently than an ordinary mathematician. For example, he recognizes that the outside of a figure is just as important as the inside, and that the kind of movement possible between outside and inside reveals something important about the mathematical nature of the figure. Also, since a topologist loves to change one figure into another, he knows that a figure's inside and outside may exchange places and become an interesting *variant* in understanding transformations.

These two puzzles show a simple way to become more aware of outside and inside.

A fly lands inside each of the shapes below and tries to cross each side only once in order to wind up outside the shape. You can see that it has no trouble doing this with the triangle. On which of the other shapes can he do this?

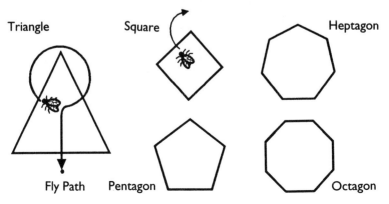

This time, the fly wants to begin and end inside the shape, crossing each side only once. You can see that he can't do it with the triangle because crossing the third side takes him outside the shape. Based on what you've concluded from the first puzzle, what sorts of shape do you think the fly needs?

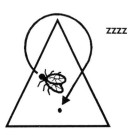

This maze makes inside and outside a little harder to tell apart. Can you tell at a glance whether each number points to the inside (enclosed with walls) or outside (open corridors or atriums) of the maze? After wandering in and out of simple geometrical shapes all day, the fly might know. . . .

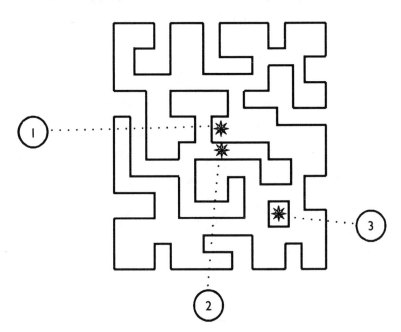

POWER LINES

Each of the five houses below needs a connection to each of the five power stations. Each house must have a power line leading to its own power station, and none of the power lines must cross. Can you draw the arrangement of power lines?

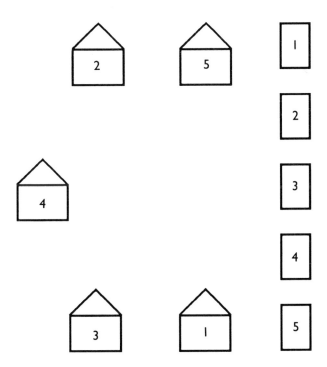

THE BRIDGES OF KÖNIGSBERG

This 200-year-old puzzle is one of the most famous topological problems of all time. It encouraged popular interest in topological map-making and led to further development of this branch of mathematics.

The 18th century town of Königsberg (also known as Kaliningrad), East Prussia, was built over the river Pregel and connected by seven bridges. The river and bridges split the town into four parts (see illustration on next page).

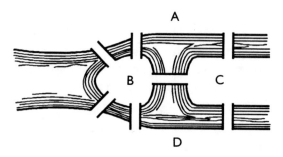

During the summer, the people of the town liked to take evening strolls across the seven bridges. But they found that in order to cross all the bridges once, they had to cross at least one bridge twice. Is it possible to cross all seven bridges just once without backtracking? The problem reached the great Swiss mathematician Leonard Euler. He studied a map of the town and transformed it into a new shape, called a *network*.

In Euler's network diagram of Königsberg, the land areas are reduced to dots and the bridges reduced to lines. Then he could more carefully reach the root of the problem.

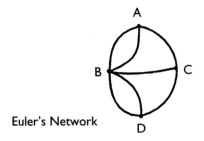

Euler's Network

See if you can retrace the lines of Euler's network without going over any line twice and without lifting your pencil from the page. What do you think Euler discovered by studying the dots and lines of his network?

Chapter 5

MECHANICAL APTITUDE

**Cranking Gears • Tugging Tugboat • Strange Shafts •
Pulleys & Ropes • Wheelies • Counterclockwise Cogs •
Gnashing of Teeth • Whizzing Water • Square Dance**

Mechanical aptitude is the ability to imagine—from nothing more than a drawing or description—certain physical properties of an object, such as how it is constructed or how it moves. It means seeing how objects fit together or move with other objects. Someone with well-developed mechanical aptitude can understand, from an engineer's blueprint, how structures appear or how a piece of machinery operates. And although mechanical aptitude problems require no mathematical figuring, they do require a natural knack for thinking mathematically.

CRANKING GEARS

If you could really turn the handle on the first gear in the direction of the arrow, do you think the lid of the box would go up or down?

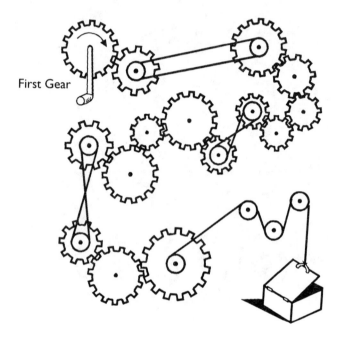

First Gear

TUGGING TUGBOAT

Which three boats will move first when the tugboat starts moving forward?

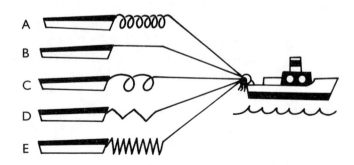

STRANGE SHAFTS

Which of the plates on the shaft (A, B, C, D, E) lift upward more than once at just one clockwise turn of the revolving shaft?

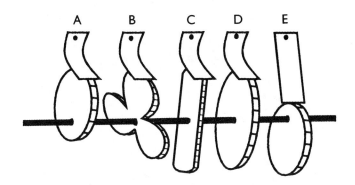

PULLEYS & ROPES

In which direction do parts of the rope A, B, C, D, and E move if you pull the rope part a in the direction indicated by the arrow?

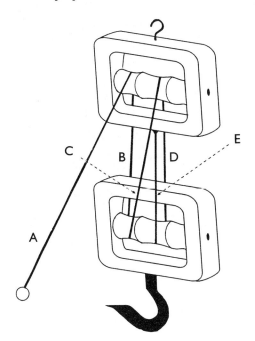

WHEELIES

Which of the wheels turns around the fastest?

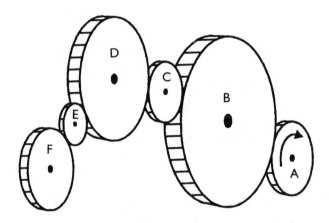

COUNTERCLOCKWISE COGS

Which cogs (B, C, D, E, F) rotate in a counterclockwise direction when cog A turns in a clockwise direction?

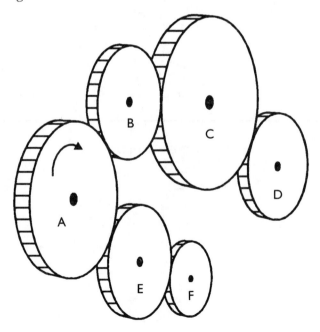

GNASHING OF TEETH

If the toothed gear A makes 14 revolutions, how many revolutions will the toothed gear B make?

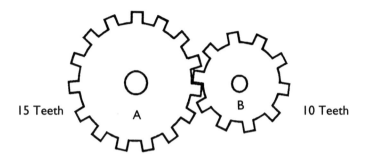

15 Teeth   A   B   10 Teeth

WHIZZING WATER

If a rapid flow of water comes through the large pipe in the direction indicated, what will happen in the smaller pipes A and B?

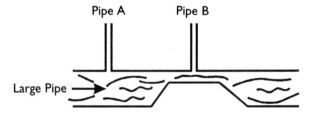

Pipe A   Pipe B

Large Pipe →

SQUARE DANCE

If the square gear A turns at 30 revolutions per minutes (rpm), how many RPMs will the square gear B complete?

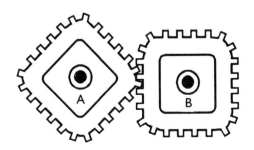

Chapter 6

LOGIC PUZZLES

**Tree-Chopping Contest ● Filled Glasses ●
Rare-Coin Thief ● Fast Fishin' ● Apricot Jam ●
The Lumberjack's Brother ● Chasing Shadows**

At the root of all mathematical problem-solving is logical thinking. In fact, learning to think logically will help you solve nearly every kind of problem, mathematical or otherwise. Logical thinking begins with careful observation of the evidence, looking for consistent and inconsistent details, and then making a series of deductions that suggest a solution.

TREE-CHOPPING CONTEST

There was a race between six tree choppers to see who could chop down a tree first. Study the drawing below. Can you tell which chopper won first place in the contest? Which choppers won second, third, fourth, and fifth places? And finally, who came in last?

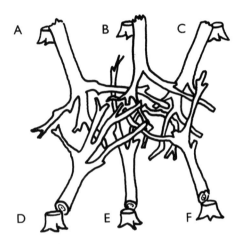

FILLED GLASSES

Of the six glasses below, three are filled with cranberry juice. By moving only one glass in the top row, make the top row resemble the bottom row.

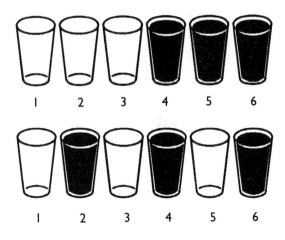

RARE-COIN THIEF

A shoplifter in a rare coin shop stole the oldest coin he could find, dated 260 B.C. If a rare coin is worth $30 for each year before Christ that it was minted, how much could the shoplifter get for his coin?

FAST FISHIN'

If five fishermen catch five fish in five minutes, how long will it take fifty fishermen to catch fifty fish?

APRICOT JAM

After picking some apricots from your tree, you decide to make some delicious apricot jam. You cut up 10 pounds (4.5 kg) of apricots, blend them together, and place them on the stove. But suddenly you remember that you were supposed to add 1 teaspoon of lemon juice for every dozen apricots. Since you can no longer count the number of apricots in your mixture, how will you know how much lemon to add?

THE LUMBERJACK'S BROTHER

A lumberjack's brother died and left a million dollars to his only brother. However, the lumberjack never received any of the money even though it was legally paid out. How could this happen?

CHASING SHADOWS

Logical problem-solving always means a careful observation of the evidence. Look at the illustration of a tall tree seen from above and the shadow it casts at various times of day. Study the shadows carefully and identify four mistakes in the picture.

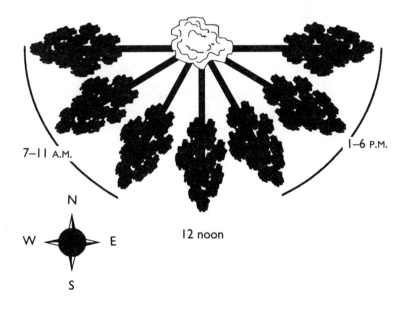

7–11 A.M.

1–6 P.M.

12 noon

N
W   E
S

Chapter 7

AMAZING MAPS & MAZES

**Canals of Mars • Tangle Teasers • Falling Rocks •
Maze Master • Rooms in the Castles**

Like everyone else, mathematicians love a challenge. Cleverly designed maps and mazes can tickle even the most serious mathematician's fancy because they provide surface models of ideas, like *fixed geometry, continuous networks,* and *topological space,* that fascinate most mathematical minds. Some of the mazes in this chapter reveal trick solutions to seemingly impossible situations. Others challenge your powers of keen observation and accurate deduction. All are fun!

CANALS OF MARS

During the 19th century, an Italian astronomer observing the planet Mars thought he saw a series of mysterious canals linked by green oases. The idea of canals on Mars became very popular, but it was also ridiculed by some scientists who claimed that what the astronomer saw was merely a figment of his imagination. Arguments went back and forth in newspapers, and in this spirit of debate, the American puzzlemaker Sam Loyd created this Martian puzzle.

Starting at the South Pole, see if you can spell out a secret message by following a canal to each lettered oasis. Pass through an oasis only once, then return to the South Pole.

North Pole

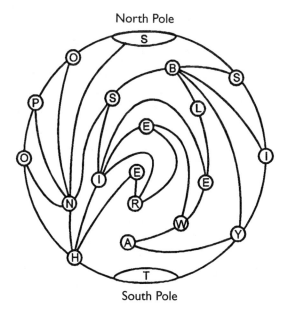

South Pole

TANGLE TEASERS

The following set of five mazes challenge you to test your speed and accuracy against various tangled-line formations. In each box you must isolate and identify one type of line from many similar lines and separate foreground from background. Mazes like these help you develop sharp linear and spatial perceptions.

As the first maze demonstrates, each line begins at a number and connects to a letter on the opposite side. Start in box 1 and move down the numbered list to 10, calling out each number and its associated letter. In Maze I below, see how box 2 connects with box E.

Maze I

Have a friend time you as you complete each box. If you can complete the whole set of mazes in five minutes or less, your line and space recognition skills are excellent.

Maze II

Maze III

Maze IV

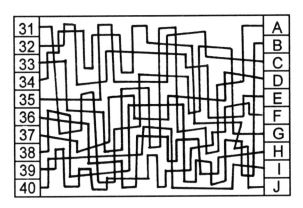

Maze V

FALLING ROCKS

The map below shows mountain roads connecting the rural town of Alpine to the city of Descanso. You can see the intersections at A, B, and C. Heavy rains have caused rockslides over some of the roads. You need to clear just one rockslide to make the shortest path between Alpine and Descanso. Which rockslide must you clear, and how many kilometers will you travel? (You can do this puzzle in miles, if you wish. Just assume all labels are in miles.)

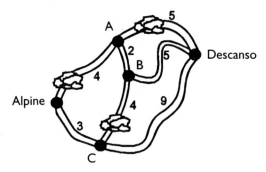

MAZE MASTER

By trial and error, you can eventually find your way out of this maze with nothing more than a pencil and eraser. But suppose this were a maze of tall hedges and you were forbidden by the Queen of Hearts to take one wrong turn or it was off with your head! Do you think you could do it?

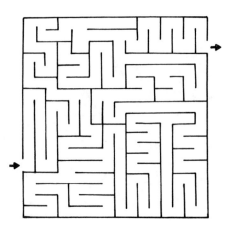

A true maze master will discover a foolproof method for getting through any maze without a single wrong turn. Are you a maze master?

ROOMS IN THE CASTLES

Here are eleven castles that you must pass through. For each, enter at the left, visit each room only one time, and exit to the right. You might find such an uninterrupted tour impossible for some castles, and a quick calculating trick beforehand will keep you from wasting steps. Do you know the trick?

Nine-Room Castles

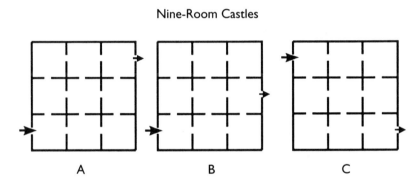

A   B   C

Sixteen-Room Castles

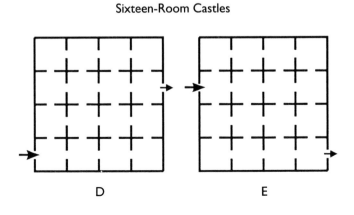

D   E

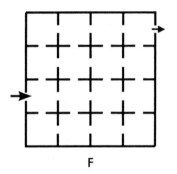

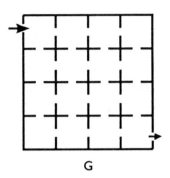

F   G

Twenty-Five-Room Castles

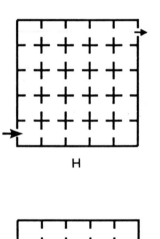

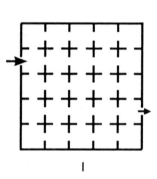

H   I

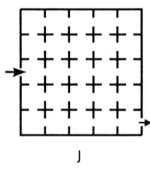

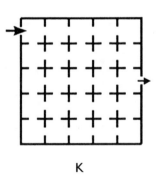

J   K

Chapter 8

FANCY FIGURING

**Bull's-Eye ● Jawbreakers ● Antsy Ant ●
A Burned Receipt ● In the Old Cemetery ●
Magic Squares ● More Magic Shapes ●
Ben Franklin's Wheel ● The Chimes of Big Ben ●
Count the Streetlights ● Weighing In**

Get out your pencil. This chapter provides some old-fashioned arithmetic problems that call for serious ciphering. Now you can enjoy an assortment of brain-teasers that touch on such weighty mathematical concerns as *probability, recombinant shapes, number series,* and some just plain count-your-fingers figuring!

BULL'S-EYE

With your bow and arrow, shoot the following scores on the target using the smallest number of arrows: A. Shoot a 25. B. Shoot a 19. C. Shoot a 47.

JAWBREAKERS

Two jawbreaker-lovers stand at the vending machine with their pennies. The machine has 35 yellow jawbreakers and 35 blue jawbreakers. If they got a jawbreaker for each penny they put into the machine, how many pennies would they use before they had two jawbreakers of the same color?

ANTSY ANT

An ant decides to race along a ruler, starting at the 12-inch end. He runs from the 12-inch mark to the 6-inch mark in 12 seconds. How many seconds will it take him to reach the 1-inch mark?

A BURNED RECEIPT

This important receipt was badly burned in a fire. Can you reconstruct the missing digits so that the equation works?

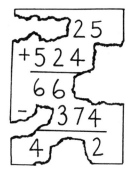

IN THE OLD CEMETERY

In the old cemetery, you stumble upon two tombstones. The dots indicate places where the stones are so worn that the dates have been erased. You find an old diary in the church and learn that both Mary and her brother John died in childhood and that the single missing digit in the bottom line of John's stone was one less than the single digit on Mary's. However, John lived *longer* than Mary. Calculate a possible birth year on Mary's stone.

MAGIC SQUARES

This ancient mathematical curiosity was once used as a charm to bring good fortune and protect against disease. Mathematicians now call it the *magic square*. It's magical because the square is constructed so that the numbers in each vertical column, horizontal row, and diagonal add up to the same number.

8	3	4
1	5	9
6	7	2

Third-Order Magic Square

16	3	2	13
5	10	11	8
9	6	7	12
4	15	14	1

Fourth-Order Magic Square

Magic squares can be of any order, beginning with a 3 × 3 square of nine numbers. In fact, any regular number sequence that you can place in a square (3 × 3, 4 × 4, 5 × 5, etc.) can be made into a magic square. The *magic constant* for a third-order square is 15. A fourth-order square has a constant of 34.

There are seven variations of a third-order (3 × 3) magic square, 880 variations of a fourth-order square, and over a million variations for a fifth-order square. Although it might seem difficult to construct anything larger than a third-order square, a French mathematician named Loubère came up with a trick for making any size of odd-order square. We'll use his method to build a simple fifth-order square.

A piece of large-grid graph paper comes in handy, since you'll be working on several squares at once, each one duplicating the original square.

Outline four neighboring 5 × 5 squares on your graph paper. You'll begin in the bottom left square. Write the number 1 in the top middle cell of that square. Write the number 2 diagonally above it to the right (in a new square), and repeat the number 2 at the bottom of that row (in the original square). Now write the number 3 diagonally to the right of the number 2, and repeat the number 3 above in the new square. Write the number 4 diagonally to the right of the number 3, and notice that another number 4 should appear in the first row of the original square, three cells from the top. When a cell is already occupied, place the next number in the cell directly below its preceding number, as in number 6.

Creating a Fifth-Order Magic Square

You should now begin to see a pattern emerge. Continue filling in the cells of your magic squares until you complete them. What is the magic constant for a fifth-order square?

17	24	1	8	15
23	5	7	14	16
4	6	13	20	22
10	12	19	21	3
11	18	25	2	9

Fifth-Order Magic Square

For a simple fifth-order square of this type, notice that the numbers proceed consecutively along the diagonal as the original square is duplicated.

You can perform all sorts of tricks with magic squares. Have a friend erase any number or series of numbers in the square and you can always replace them because you know the square's constant. Or have a friend completely rearrange all the numbers in a single row, column, or diagonal. Only you know the secret.

MORE MAGIC SHAPES

Similar to the magic square puzzles, these shapes rely on overlapping number relationships for their unusual qualities.

For the magic triangle, arrange the numbers 4 through 9 in the circles so that every side of the triangle equals 21.

For the magic daisy, place the numbers 1 through 11 in the circles so that every straight line of three circles totals 18.

For the magic star, place the numbers 1 through 12 in the circles so that each straight row of four circles adds up to 26. In this case, you may find more than one solution.

Magic Triangle

Magic Daisy

Magic Star

BEN FRANKLIN'S WHEEL

Benjamin Franklin, the great American inventor, scientist, philosopher, and statesman, was also a maker of magic squares. He preferred to design them as wheels, however, and the one reproduced below represents one of his more elaborate designs. It was supposedly doodled one tedious afternoon when the young Franklin was a clerk at the Pennsylvania Assembly.

Knowing what you now know about magic squares, fill in all the missing numbers.

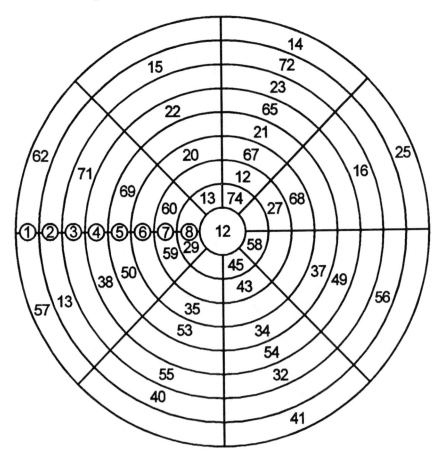

Magic Wheel

THE CHIMES OF BIG BEN

Big Ben, London's largest clock, calls out the time with loud chimes, one for each hour. If it takes Big Ben 3 seconds to chime three times at three o'clock, how long will it take Big Ben to chime six times at 6 o'clock?

COUNT THE STREETLIGHTS

On opposite sides of a street, there are 45 streetlights, each one at a distance of 30 yards from the other. The streetlights on one side are arranged so that each lamp fills a gap between two other streetlights on the opposite side. How long is the street?

WEIGHING IN

How many pounds does each cube, pyramid, and sphere weigh when each row has the combined weight indicated?

Chapter 9

LEWIS CARROLL, PUZZLEMAKER

**The Monkey & the Weight ● Crossing the River ●
Doublets ● Name Anagram ● Division by Nine ●
A Small Dinner Party ● Two Broken Clocks ●
Why Is a Raven Like a Writing Desk?**

The 19th century English author of Alice tales, Lewis Carroll, was also a clever puzzlemaker. Carroll's *Alice's Adventures in Wonderland* (1865) and *Through the Looking-Glass* (1872) contain some of the best riddles and puzzles ever invented. Carroll's poems, stories, and letters are filled with many original brain-teasers that have become familiar today. This shy Oxford mathematician with a stammer, Charles Ludwig Dodgson (1832–1898), who adopted the pen name Lewis Carroll, contributed to puzzle magazines regularly and also wrote serious books on math. Carroll was perhaps more admired by contemporaries for his puzzle-making and mathematical abilities than for "dabbling" in fiction.

THE MONKEY & THE WEIGHT

A rope hangs over a pulley wheel suspended high above the ground. A monkey hangs on to one side of the rope and a weight attaches to the other side of the rope, counterbalancing the monkey. If the monkey climbs the rope, what will happen to the weight opposite him? Will it   (a) go up, (b) go down, or (c) remain in the same place?

CROSSING THE RIVER

Four gentlemen and their wives wanted to cross the river in a boat that held no more than two people at a time. The conditions were: A gentleman could not leave his wife on the bank without him unless she was either alone or only in the company of one or more women, and that someone must bring the boat back. How did they do it?

DOUBLETS

In a *doublet,* you change one word into another word one letter at a time. In the process, you create linking words. The mathematical challenge is to do this in the fewest steps possible. For example:

Make BREAD from FLOUR.

FLOUR
floor
flood
blood
brood
broad
BREAD

Try these words.

Cover EYE with LID.
Send JOE to ANN.
Pluck ACORN from STALK.
After DINNER, serve COFFEE.

NAME ANAGRAM

During the Mad Hatter's tea party in *Alice's Adventures in Wonderland*, the Dormouse tells a story about a little girl who falls into a well. The girl's name is "Lacie," the name "Alice" with the letters changed around.

Lewis Carroll loved to invent similar anagrams, some of them quite complicated. He was fond of putting famous peoples' names—particularly long names—in anagram form so that complete sentences were formed.

The next anagram disguises the full name of a famous woman of Carroll's time. Can you identify her? *Hint:* She was an American.

Flit on, cheering angel!

DIVISION BY NINE

Late one sleepless night, Lewis Carroll discovered a way to divide any number by 9 without doing any division. The answer comes from adding and subtracting the digits of the number you wish to divide. If you hate long division and don't have a calculator, you may find his method useful.

To divide 1,950 by 9,

Add the digits 1 + 9 + 5 + 0 to get the sum of 15.

Add the digits of 15 to get the sum of 6, your remainder.

Place the remainder sum 6 over the digit 0 in the ones column of 1,950 and subtract.

$$
\begin{array}{r}
6 \\
\underline{1950} \\
6
\end{array}
$$

Place that difference over the digit 5 in the tens column of 1,950 and subtract again.

$$
\begin{array}{r}
66 \\
\underline{1950} \\
16
\end{array}
$$

Place that difference over the digit 9 in the hundreds column and subtract a third time. Remember to *borrow and carry*—in this case you must borrow 1 from an unknown digit in the thousands column.

$$\begin{array}{r} 166 \\ \underline{1950} \\ 216 \end{array}$$

Place that difference over the digit 1 in the thousands column and subtract one last time.

$$\begin{array}{r} 2166 \\ \underline{1950} \\ 216 \text{ R. } 6 \end{array}$$

The number 216 is the answer, with a remainder of 6, identified earlier.

This method also makes it easy for you to tell almost immediately whether a number is evenly divisible by nine. If you wind up with a remainder of nine after you add the digits and sum of digits, then the number you started with may be evenly divided by nine. Continue with all steps as before—placing the nine over the ones column, etc.—but make sure you *add one* to your total for the correct answer.

Now try this method with the numbers: 2,776, 5,678, and 3,897.

A SMALL DINNER PARTY

Here is a puzzle from Carroll's book of puzzles, riddles, and anagrams, *A Tangled Tale* (1885).

"The governor wanted to give a small dinner, and so he invited only the following guests: his father's brother-in-law, his brother's father-in-law, his father-in-law's brother, and his brother-in-law's father. How many guests came to dinner?"

TWO BROKEN CLOCKS

Which is better, a clock that is right only once a year or a clock that is right twice a day?

WHY IS A RAVEN LIKE A WRITING DESK?

This riddle, one of the most famous, is from *Alice's Adventures Underground*. Carroll later confessed that when he thought of it, he had no answer in mind but finally felt compelled to invent one because of the pressure from readers. He published this answer.

Why is a raven like a writing desk?
Because it can produce a few notes, though they are very flat.

Others have come up with clever answers to this riddle. Sam Loyd provided this solution.

Because Poe wrote on both.

Other readers have suggested these responses. Can you add your solutions to the list?

Because it slopes with a flap.
Because both have quills dipped in ink.

Answers

1. Grid & Dot Games

Answers are within the chapter.

2. Sum of the Parts

Stamp Stumper
There are eight sets, however you wish to divide them.

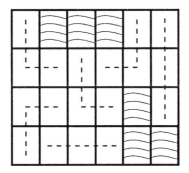

Broken Dishes
You'd have six dishes.

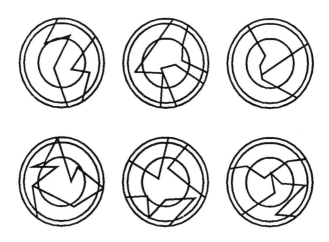

Cut the Pizza

Two cuts will make no more than four pieces. A third cut will make no more than seven pieces. So, in order to make eight pieces from three cuts, you need to think three-dimensionally and stack the four quarter-pieces on top of one another.

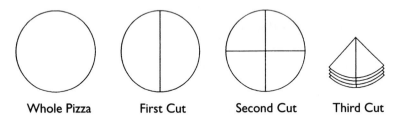

Whole Pizza   First Cut   Second Cut   Third Cut

Fractured Fractions

Since all the numbers are multiples of 4, you can solve the problem by first finding three-fourths of the number and then finding two-thirds of that answer.

For the number 12, 3/4 of 12 is 9; 2/3 of 9 is 6.
For the number 20, 3/4 of 20 is 15; 2/3 of 15 is 10.
For the number 32, 3/4 of 32 is 24; 2/3 of 24 is 16.
For the number 44, 3/4 of 44 is 33; 2/3 of 33 is 22.
For the number 52, 3/4 of 52 is 39; 2/3 of 39 is 26.

The answer will always be half the original number.

Divide the Time

When you add all the numbers on the face of the clock you get a sum of 78. Since two intersecting lines always make four sections, and since 78 cannot be divided evenly into four sections, the lines you draw must not intersect. Instead, you must draw two parallel lines that divide the clockface into three sections.

$$11 + 12 + 1 + 2 = 26$$
$$10 + 9 + 3 + 4 = 26$$
$$8 + 7 + 6 + 5 = 26$$

Parcels of Land

Divide the original square into fourths. One-fourth (square A) was willed to the landowner's wife. Since there were three squares left to be subdivided among the four sons, removing one-fourth of each remaining square leads to the arrangement below.

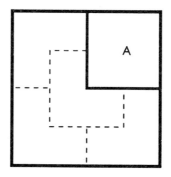

Four Lines in a Square

You can draw a maximum of eleven sections in the square using only four straight lines. To do this, each line must intersect all the other lines.

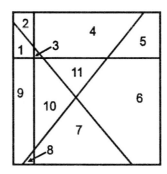

Count the Blocks
The number of blocks are: (A) 35, (B) 49, and (C) 54.

Sides, Edges & Corners
Block A touches nine other blocks, block B touches four other blocks, and block C touches 13 other blocks.

Crayon Constructions
1. How do you determine how many squares of the same size you can construct with 24 crayons?

To find out how many squares, divide 24 by one of its multiples 2, 4, 6, or 8, and then divide the result (quotient) by 4 for the total number of squares. For example:

$$24 \div 2 = 12$$
$$12 \div 4 = 3$$

Or, three squares with two crayons to a side.

$$24 \div 3 = 8$$
$$8 \div 4 = 2$$

Or, two squares with three crayons to a side.

2A. Two squares with three crayons to a side give you one smaller square.

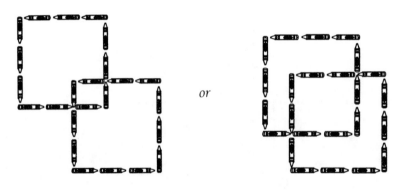

or

2B.

2C.

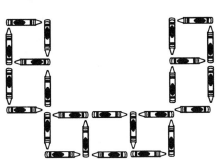

2D. One larger square.

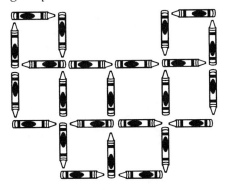

Two larger squares.

Five larger squares.

In all cases, the correct answer depends on having the minimum number of crayons on the perimeter. The most economical area is the one with the smallest perimeter.

Box the Dots
Here's how you can box the dots.

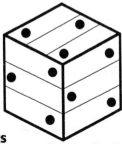

3. Vanishing Tricks

The Lost Line
The tenth line was "absorbed" into the nine lines so that you can no longer see it. If you measure one of the nine lines and then measure one of the original ten lines, you will discover that each of the nine lines is slightly longer than each of the ten lines.

Disappearing Square
Remember, the trick lies in the way you drew the diagonal line. Because that line does not equally divide the paper from corner to corner, the half-square you snipped off is taller than 1 inch (1⅓

inches). This means that the rectangle's height is actually 5⅓ inches, not 5 inches, and that the vanished square has been absorbed into this new area.

Where's the Wabbit?

The eleven rabbits have turned into ten rabbits. Like the Vanishing Line and Vanishing Square tricks, the 11th rabbit has been absorbed into the recombined set of 10 rabbits, making each new rabbit just a little larger than each rabbit of the original set.

The Floating Hat

Although you lose one head when you recombine them, you still have six hats. Examine your new heads against the old ones. You will find that each new head is larger than each original head, although the size of the hats remains the same.

Lions & Hunters

Although you start with seven lions and seven hunters, when you turn the circle to the left you wind up with eight lions and six hunters. The area of each lion and hunter has been broken up and redistributed into this new arrangement of figures. Notice how each of the eight lions is smaller than each of the original seven lions and how each of the six hunters is larger than each of the original seven hunters.

Funny Money

In redistributing the area of seven bills into eight bills, each of the eight bills will now be one-seventh shorter than each of the original seven bills.

Mysterious Tangrams

Here's how the animals are constructed.

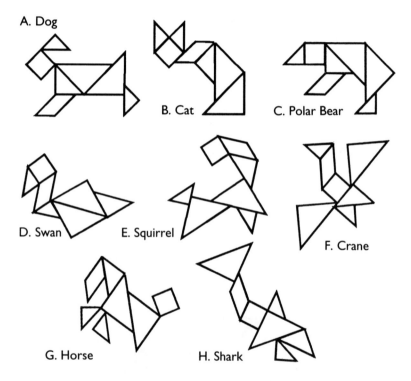

A. Dog

B. Cat

C. Polar Bear

D. Swan   E. Squirrel

F. Crane

G. Horse   H. Shark

Tangram Oddities

Consider these oddities. If you carefully compare the tangrams in each of the two sets, you will see that they differ in size. Start with the pigs. Remove the tail from the pig at the right and then compare the bodies of the pigs. The pig with its tail removed is smaller in area, though you won't notice that so easily with the tail attached.

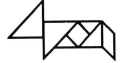 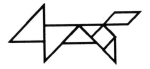

Now remove the foot from the human figure on the right and compare the figures. Notice how much thinner the footless figure on the left appears.

You've seen the same mathematical principle at work in previous puzzles: concealed distribution. Although the tangrams appear identical except for the extra piece, their dimensions differ and their sameness is an illusion.

4. Stretchy Shapes & Squiggly Lines

A Fly's View

To wind up outside the shape, the fly needs shapes with an odd number of sides—the triangle, the pentagon, and the seven-sided shape (heptagon). Since the fly begins inside, it has to cross an odd number of sides to end up outside.

The fly needs shapes having an even number of sides—the square, the hexagon, and so on, to end up inside the shape.

For the maze, count the number of lines (walls) you cross as you follow each dotted line out of the maze. If you cross an *odd* number of walls, you are in an enclosed area of the maze; if you cross an *even* number of walls, you are in an open corridor.

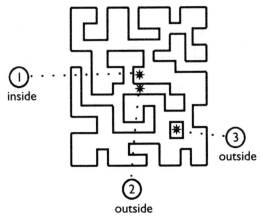

Power Lines

You can't solve a topological problem like this by using straight lines. Topological reasoning separates the interior and exterior regions of any object so that lines must pass around the object to reach their destination.

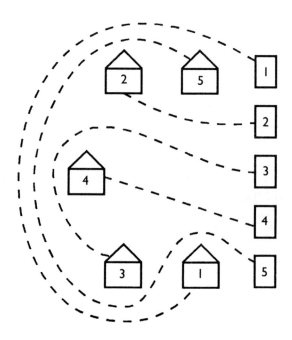

The Bridges of Königsberg

Euler discovered that the town of Königsberg needed one more bridge for people to cross each bridge once without backtracking. With another bridge, Königsberg would become a *one-stroke network*, that is, a network that you can trace in one continuous line.

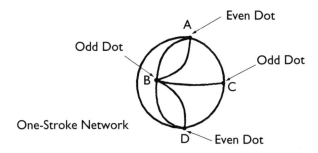

Euler's simple method for recognizing one-stroke networks was to first count the number of lines leading to each dot and then call each dot "odd" or "even" accordingly. By trial and error, he discovered that any network can only have *an even number of odd dots, or no odd dots at all.*

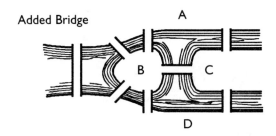

When a network had two odd dots, it was a one-stroker, provided you began at one of the odd dots. But if a network, like Königsberg, had four odd dots—or six, or eight, or *any* even number other than two—it was not a one-stroker. Euler suggested building another bridge on the west side of the city to connect opposite shores. In doing so, the network was changed as shown.

5. Mechanical Aptitude

Cranking Gears

Up. Meshed gears turn in opposite directions, as do gears attached by a twisted belt. Gears attached by an untwisted belt turn in the same direction. The sum of rotation for the assembly of gears results in the last gear rotating away from the box, pulling the string into the gear, and so lifting the lid of the box.

Tugging Tugboat

The first three boats to move are B, C, and D.

Strange Shafts

The plates turning upward more than once are B and C.

Pulleys & Ropes

Rope parts B and D move up, A and C move downward, and E stays the same but grows shorter as the bottom pulley moves toward the top pulley.

Wheelies

The wheel E. The smallest wheel will always rotate the fastest.

Counterclockwise Cogs

Cogs B, D, and E move counterclockwise.

Gnashing of Teeth

The second gear (B) will make 21 revolutions.

Whizzing Water

Since the larger pipe narrows before the opening of pipe B, water pressure will increase at that point and force the water up through pipe A.

Square Dance
B will complete 30 RPMs. Both gears will turn at the same speed.

6. Logic Puzzles

Tree-Chopping Contest
In first place was E; second place, C; third place, A; fourth place, B; fifth place, F; and last place, D.

Filled Glasses
A common-sense solution is in order here: pour the cranberry juice in glass #5 into glass #2 and return empty glass #5 to its original position.

Rare-Coin Thief
Zero. The coin is counterfeit because the term B.C. ("Before Christ") makes no sense on an ancient coin. How could the minter know that Christ would be born 260 years *after* he minted the coin?

Fast Fishin'
Five minutes. Each fisherman takes five minutes to catch a fish no matter how many fishermen are fishing. The time element remains constant and does not influence the outcome of this problem.

Apricot Jam
Count the pits.

The Lumberjack's Brother
The lumberjack was female.

Difficulties in figuring this one out have to do with making a false connection between gender and an appropriate job. This leads to the mistaken assumption that all lumberjacks must be men.

Chasing Shadows

1. The sun rises in the east, so the 7 A.M. to 11 A.M. morning shadows would be to the left of the tree, not to the right.
2. At 12 noon there would be no shadow since the sun is directly overhead.
3. The sun sets in the west, so the late afternoon and evening shadows would be to the right of the tree, not to the left.
4. The shadows should not all be the same length, since they shorten as the day approaches 12 noon, then lengthen again in the late afternoon and evening.

7. Amazing Maps & Mazes

Canals of Mars

When this puzzle first appeared in a magazine, over 50,000 readers reported, "There is no possible way." But the puzzle is really very easy to solve.

The oases connect to spell out: "THERE IS NO POSSIBLE WAY."

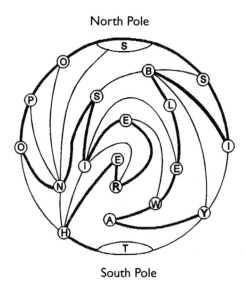

North Pole

South Pole

Falling Rocks

Clear the rockslide on the road between Alpine and intersection A, continue on the road to intersection B, then go from B to Descanso. The trip totals 11 kilometers. (Or 11 miles if you assumed all labels were in miles.)

However, by driving an extra mile through C to Descanso, you'd probably save time, since it would surely take several minutes to clear a rockslide.

Tangle Teasers

Hint: trace these lines in different colors.

Maze I. 1-C, 2-E, 3-H, 4-F, 5-G, 6-B, 7-A, 8-D, 9-J, 10-I.
Maze II. 11-H, 12-G, 13-C, 14-D, 15-A, 16-I, 17-J, 18-E, 19-B, 20-F.
Maze III. 21-E, 22-C, 23-H, 24-D, 25-I, 26-G, 27-A, 28-F, 29-J, 30-B.
Maze IV. 31-G, 32-E, 33-A, 34-J, 35-C, 36-H, 37-D, 38-B, 39-F, 40-I.
Maze V. 41-I, 42-E, 43-F, 44-G, 45-C, 46-J, 47-H, 48-A, 49-B, 50-D.

Maze Master

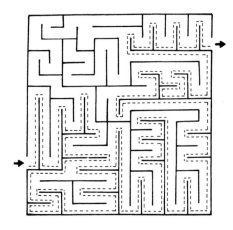

To prevent yourself from backtracking or going down blind corridors, keep one hand, the left or the right, on the wall from the moment you enter the maze until the moment you exit. Although choosing one hand rather than another may take you on a longer route, either hand will lead you out.

Rooms in the Castles
Castles B, G, and I are impossible to tour without backtracking. Here are routes for visiting the other castles. Alternate routes are possible.

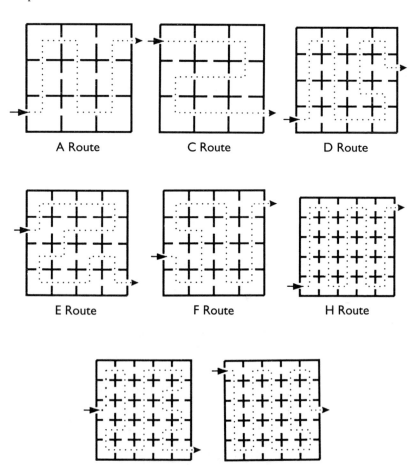

A Route   C Route   D Route

E Route   F Route   H Route

J Route   K Route

Can you see the pattern? To determine whether or not you can pass through a castle's rooms only once, first count and number all the rooms, beginning from the room at the top left-hand corner.

1	6	7
2	5	8
3	4	9

If the castle has an odd number of rooms, you must enter an odd-numbered room and exit through an odd-numbered room. You cannot enter or exit through an even-numbered room.

If the castle has an even number of rooms, you must either enter an odd-numbered room and exit through an even-numbered room, or enter an even-numbered room and exit through an odd-numbered room.

Also, if you can pass through a castle without repeating or missing rooms, then you can do it several different ways.

You can now see how each solution to the castle puzzles sticks to this formula: A. 3 to 7, B. 3 to 8, C. 1 to 9, D. 4 to 15, E. 2 to 13, F. 3 to 16, G. 1 to 13, H. 5 to 21, I. 2 to 24, J. 3 to 25, and K. 1 to 23.

8. Fancy Figuring

Bull's-Eye

A. 16, 8, 1 (three arrows); B. 16, 2, 1 (three arrows); and C. 32, 8, 4, 2, 1 (five arrows).

You can form any whole number up to 63 on this target. Each number is a power of 2, which means that you can create any other whole number by combining numbers.

Jawbreakers

Three pennies. After the second penny, they would have either two yellow or two blue jawbreakers, or a yellow and a blue jawbreaker. A third penny would deliver a jawbreaker that had to match one of the colors.

Antsy Ant

Ten seconds. Since it takes the ant 12 seconds to cover the distance between 12 and 6 inches, it takes him 2 seconds to travel each inch. You can divide that distance into six 2-second time intervals.

Since it's a shorter distance between 6 inches and the 1-inch mark, you can divide that distance into only five 2-second intervals. So it takes the ant only 10 seconds to cover the remaining distance.

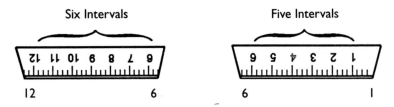

Six Intervals	Five Intervals
12 6	6 1

A Burned Receipt

$$
\begin{array}{r}
1425 \\
+5241 \\
\hline
6666 \\
-2374 \\
\hline
4292
\end{array}
$$

In the Old Cemetery

Mary was born in 1896, 1897, 1898, or 1899.

For example, if Mary were born January 5, 1897, and died on her birthday, January 5, 1903, she would have died on the first day of her seventh year. The year 1900 was not a leap year, since it was centesimal (ending in 00); so, there were no leap years in Mary's lifetime. Thus, she lived exactly six years of 365 days each, or 2,190 days.

If her brother John were born January 5, 1903, and died the day before his birthday, January 4, 1909, it would have only been the last day of his *sixth* year. However, during John's lifetime there would have been two leap years, 1904 and 1908. Thus, although he lived six years minus one day, two years had an extra day, making his lifetime a total of 2,191 days—one day longer than Mary's.

For this to be possible, Mary must have been born no earlier than March 1, 1896, since her last year was expressed in a single digit.

Magic Squares
The magic constant for a fifth-order square is 65.

More Magic Shapes

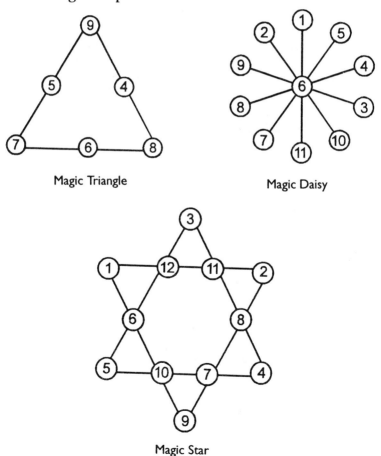

Magic Triangle

Magic Daisy

Magic Star

Ben Franklin's Wheel

Missing numbers, clockwise starting above the circled band numbers are: Band 1: 73, 30, 46   Band 2: 24, 47, 63   Band 3: 64, 39, 48   Band 4: 17, 70, 33   Band 5: 66, 18   Band 6: 19, 51, 36, 52   Band 7: 75, 28, 44   Band 8: 26, 61, 42.

The magic number that the rings add up to is 360, the number of degrees in a circle.

The Chimes of Big Ben

It takes 7½ seconds. You can illustrate the chiming of Big Ben like this.

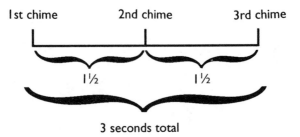

3 seconds total

The first chime accounts for the interval from the first to the second chime. The second chime accounts for the interval from the second to the third chime and ends with the third chime. Therefore, you must divide the 3 seconds in half since only two intervals exist for three chimes, averaging 1½ seconds per interval.

At 6 o'clock, you would have five intervals totalling 7½ seconds as shown.

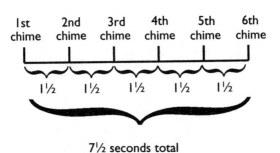

7½ seconds total

Count the Streetlights

It's 660 yards. There are 23 lamps on one side and 22 on the other side. There are 22 gaps between 23 lamps; therefore, the street is 22 times 30 yards long, or 660 yards.

Weighing In

The cube is 6 pounds, the pyramid 3 pounds, and the sphere 2 pounds.

9. Lewis Carroll, Puzzlemaker

The Monkey & the Weight

The weight and monkey would always remain opposite each other. As the monkey climbs, the weight moves up.

Crossing the River

Here's a code for the four gentlemen and their wives. M1 and W1 for the first man and first wife, M2 and W2 for the second man and second wife, M3 and W3 for the third, and M4 and W4 for the fourth.

1st crossing: M1 and W1 cross; M1 returns.
2nd crossing: M2 and W2 cross; M2 returns.
3rd crossing: M1 and M2 cross; M2 and W2 return.
4th crossing: W2 and W3 cross; M1 returns.
5th crossing: M1 and M2 cross; W3 returns.
6th crossing: M3 and M4 cross; M3 returns.
7th crossing: M3 and W3 cross; M4 returns.
8th crossing: M4 and W4 cross.

Name Anagram

Florence Nightingale

Doublets

EYE	JOE	STALK	DINNER
lye	doe	stale	sinner
lie	die	stare	singer
LID	did	scare	linger
	aid	score	longer
	and	scorn	conger
	ANN	ACORN	confer
			coffer
			COFFEE

Division by Nine
308 remainder 4; 630 remainder 8; and 433.

A Small Dinner Party
One.

Two Broken Clocks
This is Lewis Carroll's answer to the problem.

   "I have two clocks: one doesn't go at all, and the other loses a minute every day; which would you prefer?

   " 'The latter,' you reply, 'unquestionably.'

   "Now attend: The one which loses a minute a day has to lose twelve hours or seven hundred and twenty minutes, before it is right again, consequently it is only right once in two years, whereas the other is evidently right as often as the time it points to comes around, which happens to be twice a day."

Index

Page numbers in italics are answers.